LIFE SKILLS

IN THE PACIFIC

Basic Building

Eron Hagunama

OXFORD

Contents

Introduction

Basic Building is the companion book to *Basic Maintenance*. It is a step-by-step instructional book about:

- safe building practices
- the use of tools used for building
- the preparation of traditional building materials
- the construction of simple building structures
- the use of common bush materials to build traditional houses in Papua New Guinea.

This book is written to complement the Upper Primary Syllabus subject 'Making a Living', and supports the philosophy of education, which emphasises relevance and self-reliance. *Basic Building* teaches students useful skills and provides essential knowledge to enable them to become useful members of their local communities.

Strand: *Better Living*
Sub Strand: *Care and management*
Outcomes: 7.2.2 *Assess home and school buildings to identify areas that require maintenance, repair or other improvements and undertake appropriate actions*
8.2.2 *Work collaboratively with others to select and undertake a project based in identified needs within the school or community.*

I encourage students not only to read this book but also to use the information in this book to develop useful skills that will assist them in their life.

Eron Hagunama

Building in Papua New Guinea

Houses and other structures can be built from a wide range of building materials, such as timber, **grass**, **bamboo**, bricks and cement. Traditional houses in Papua New Guinea are built from materials that are readily available to everyone, which means that the only costs association with constructing a new home come from labour. Such houses are built from local materials found in the immediate or nearby area and they are built in a traditional way for each region.

Our traditional houses also serve the specific requirements of our climate. This means that houses near the water or in areas that have regular floods are built on high posts to keep the living environment dry and safe. Houses in higher regions, where temperatures can be cold, are built close to the ground so that heat is retained and residents are kept warm. Our haus win allow for the free circulation of air, which is important in hot climates as enclosed houses can become very hot and stuffy if fresh air is not able to enter easily.

Many people in Papua New Guinea would like to live in a more western-style house. But modern housing is not always suited to our climate. More importantly, they cost more money to build, as building materials are not readily available the way that grasses, local timber, mud and leaves are. When considering which type of house you would like to live in, you should consider some of the advantages and disadvantages of traditional and western houses.

Traditional house	Western house
• does not last long	• lasts for a longer time
• more labour intensive	• less labour intensive
• doesn't have a roof suitable for collecting water	• more expensive to build
• much cheaper than western style houses	• poor insulation
• better insulated	• corrugated roof very noisy when it rains
• less noisy in the rain	
• gives a sense of belonging to the community	

Village houses in Papua New Guinea

Highlands houses

Most highlands houses are built on the ground so that the cold air does not blow through the house. The floors are dug into the ground slightly, to a depth of about 30 cm, to keep the house warm. Whatever the shapes of the houses are, all houses are completely enclosed with only one door. House posts are closely planted with rolls of bamboo or **blinds** spread against posts inside the house. There is always a fireplace dug inside the house. Floors are covered with bamboo mat and all roofs are thatched.

Southern Highland houses

People in the Southern Highlands build small round houses, which usually have a kunai thatched roof. Flattened and woven blinds are used as walls. Sewn pandanus leaves are used for walls or for roofs.

Eastern Highland houses

People in the Eastern Highlands live close to coastal land. Their houses are either round or rectangular. Both types of house are built directly on the ground and without windows. Bush materials, such as split bush timber, are traditionally used.

Western Highland houses

In the Western Highlands, people build spacious round houses that are large enough for extended families. The houses are enclosed by split timber poles that have been cut to the same height and sharpened at both ends. Usually **pitpit** or bamboo blinds are used to make the inside walls. Long split bush timber is used to provide a framework for the roofs. Strong pitpits are used as supporting materials to hold the **rafters**. The roofs themselves are made of thatched kunai. A single door ensures that the house is kept warm.

kunai grasses

rafters

inside wall

Coastal houses

Generally, most coastal houses are built on posts. The houses are built on posts that are high enough for cool air to flow through, as it is very hot in the coastal areas. The houses are lightly enclosed with windows in each room. Spacious verandahs allow for large family gatherings and relaxation.

Markham houses

Markham village house are built on bush timber posts that are about one metre high. These round houses are built high enough to prevent water from entering the house during the wet seasons. Kunai grass is used for thatching, and other common building materials are bush timber, split coconut stems, bamboo, pitpits, blinds and strong bush rope.

Madang coastal houses

People living along the coast of Madang province build their houses on strong bush timber posts or milled timber posts. Roofs are covered with **morota** thatching and walls are lined with **sago palm** wall palings. Split palm stems or flattened bamboos are used as flooring. These houses usually have many windows, spacious rooms and large, extended verandahs.

Sepik River houses

The people living along the Sepik River build their houses along the banks of the river. During the wet seasons, the river rises and floods. Therefore, the houses in this region are built on high, strong **kwila** posts to keep the house and its contents dry and safe. Split coconut stem and bark of palm trees are used for flooring. Neatly lined sago palm leaves or sewn bamboo splints are used for walls. Bush timber and strong bamboos are used for the framework of the building and the roofs are thatched with morota. Strong bush ropes and cane splints are used to fasten the joints and materials firmly.

Tapini houses

The Central Province people living along the Kokoda trail have a different style of house. Tapini houses have high pitched roofs that slope towards the ground. The roof rafter framework also serves as the walls of the house. These houses are built quite small to keep the family warm.

Trobriand Island houses

People living in the beautiful Trobriand Islands in Milne Bay Province build their houses on high kwila posts. The kwila posts last for a long time and so can be used several times to rebuild the house. The space under the floor is used to store firewood and foods, such as yam, banana and taro. The floor is made of split coconut stems and the walls are made of woven coconut leaves. The roof is covered with pandanus leaves, neatly folded over bamboo splints. Windows allow cool sea breezes to flow through the house. Cooking can be done on the veranda or in the house.

Gogodala houses

People from the Western Province build their houses on large, high kwila posts. Firm ground is chosen on which to build their houses, as most of the land is swampy. The house posts are high enough to allow people to cook food under the house. Another reason for building the house so high is that the wind will blow away mosquitoes. The floor is made from split and flattened bark of palm trees. The wall frame is covered with sewn sago leaves. The roof is covered with woven and sewn sago leaves. There are windows in each room. There is usually a verandah in front of the house for cooking and family gatherings.

New Guinea Islands houses

Tolai houses

The **Tolai** people build their rectangular houses on high posts made of either matured coconut stems or tree post from a legume tree (called **luesino** in Kuanua dialect). The floor is made of split coconut stems or flattened bamboo. The wall and roof frames are made of bush timber. The walls are covered with split coconut or palm stems, while the roof is covered with kunai grass. Cooking is done in a separate haus cook built nearby.

Bougainville houses

The people of Bougainville build two types of houses. The people living inland of the island build their houses on the ground, while the people living along the coast built their houses on high posts. But the materials used on both houses are similar. The posts are made from bush trees called **tolas**. Both kinds of houses are rectangular and have triangular roofs made with kunai grass. The floors are made from the bark palm trees. The walls are covered with bamboo blinds.

Houses over the water

Most people living along coastal areas build their houses on high posts over the seawater. Houses are built closely to each other on **platforms** supported with strong **mangrove** posts. Houses built over the sea must be built with strong materials and fastened firmly against strong winds and storms. Walkways, which serve as pathways between the houses, are built from the land into the water. Houses are not completely enclosed so that cool evening sea breezes may flow through. It is not necessary to build toilet pits with such houses, as human waste can simply be dumped straight into the sea, providing turtles and fish with nutrients.

Good building practices

It is very important to make sure buildings last a long time no matter which building materials we use. This means that building materials need to be chosen very carefully and prepared properly. If the woods you choose are inappropriate for the climate or other requirements of your region, your building will not last as long. Once a building is completed, the wood used in the building should be as dry as possible to prevent rotting in the future and the eventual decay of the entire building.

The following points should be considered when selecting and preparing wood for building.

Choosing the right wood

There are many types of trees in Papua New Guinea. This means that the wood you choose for your building will be different, too, and will have different properties. Some woods are hard and durable; others are soft and thus decay easily. Woods that are sturdy enough to last a long time do not need to be treated with any kind of preservative. For woods that are less durable, it is important to properly **season** (treat) them. Before you begin to build, find out what kind of wood in your region is best suited for the task at hand.

Protecting post ends

When posts are placed in wet ground, the posts themselves will become wet and begin to rot after a while. Therefore, it is important to treat the posts with chemicals to prevent rot from setting in. If you start with dry wood, however, this is a better option. You can keep wood that is placed in holes in the ground dry by painting the ends of every post with **tar** or bitumen. This will keep wetness away from the wood, thus keeping it from rotting. If you do not have access to tar or bitumen, you can use dirty engine oil.

Protecting fence posts and outside poles

When a wooden post is placed in the ground, either to form the frame of a house or to serve as part of a fence, the top of the post will be exposed to rain. The splits in the wood allow rain water to seep through the pole and may cause the wood to rot. This can be avoided if you sharpen the top of each post to form a point at the top. Rain water will run off the sharpened slope. If you are able to paint the pointed top with tar, or put a tin or iron cap on it, it will protect the wood even more.

Keeping wall linings dry

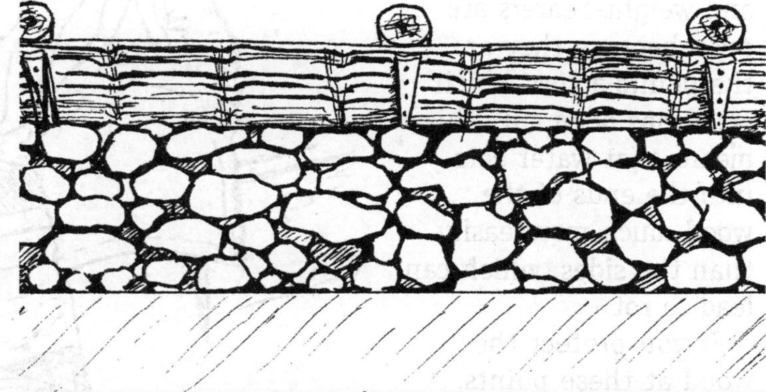

bamboo or pitpit

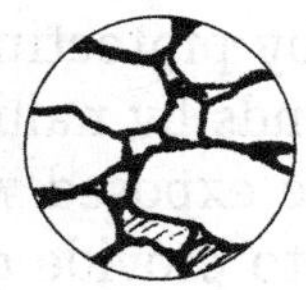

stone or brick

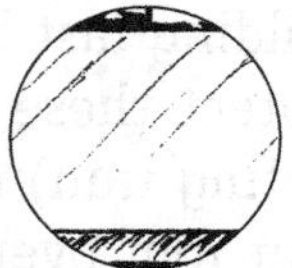

concrete slab

Wall lining should be kept clear of the ground so that they will not rot quickly. Most wall linings are made of thin materials (such as pitpit, bamboo and **weatherboard**), which will decay and rot far more quickly than large pieces of wood.

If wall linings are not close to the ground, they will dry more easily if they get wet. Keep them away from the ground so that when they do get wet they can dry off easily.

If your building has a concrete base, you can build a brick or stone wall up to a height of about 50 or 60 centimetres. You can then begin adding the thinner wall linings on top of that. If you do not have bricks or stones, you could use a treated log as a base for your wall.

Creating roof overhangs

Roof **overhangs** provide protection from the rain for the walls of your building. This means that a wider overhang will provide more protection than a narrow overhang. An overhang should be at least 60 centimetres wide. An overhang that is a metre or more wide will be even more effective.

Overhangs not only protect the building from rain, they also keep the sun off the walls, which will keep the inside of the building cooler.

Protecting the ends of wood

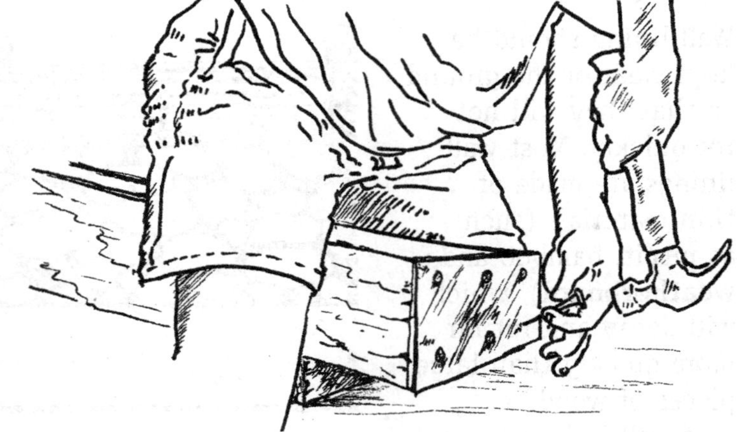

The ends of joints, rafters and weight-bearers are open because they are the points across which wood has been cut. This means that water gets into the ends of the wood much more easily than the sides, which can lead to rot.

If you protect the wood at these points, you will make your building last longer by protecting it from decay. You can protect these open ends by nailing boards or metal (such as roofing iron) over the exposed wood. You could also extend your roof overhangs to provide greater protection to the exposed wood. Painting the ends with tar or ordinary house paint will also protect the wood.

Preventing rot in flooring

When water invades wood and is left to seep in, the wood will rot and decay. This problem can be avoided if, when you build, you leave a small space (about 5 mm) between the floorboards to let the water run out. This can be done on all places where regular exposure to water is expected, such as verandahs and shower rooms.

Traditional bush building materials

There are many different types of bush materials used to build traditional village houses. Bush materials are cheap and can be easily collected from the bush. They are suited to the climate and to Papua New Guinean culture.

Type	Building uses
bamboo	house walls, house frames, wall supports, floors
tree trunk	posts where strong post is needed
bush vines	fastening or lashing together of house structure
grass	roof thatching
pandanus	walls, windows, floor coverings flatten and woven into walls
sago palm leaves	walls, roofs
river stones	walls, floors
bush timber	house posts, walls, floors, wall frames, garden fences, windows
bark	lining walls, floors and on roofs

Bamboos

Bamboos are found in all parts of Papua New Guinea. They grow well on shores, plains, forest, hillsides and mountains. The stems of bamboos are cylindrical and hollow. They are separated by partitions or joint called **nodes**. The space in between nodes is called the **internode**. There are two common types of bamboos. They are the strong bamboo and the other is the soft bamboo. The soft bamboo can be used for weaving blinds for walls while you can use the strong bamboo for flooring, house posts and roof frames.

HOW TO PREPARE BAMBOO:

1 Select and cut matured bamboo and cut off the branches. To split bamboo, start from the narrow top end and work through to the wider bottom end. Press one half on the ground firmly while lifting and opening the other half apart. Use a bush knife to cut through if the internode is strong.

2 To remove the internode, use a bush knife to remove the nodes or smash the nodes by hitting the nodes with a wooden mallet, a piece of heavy timber or an axe head

3 To flatten the bamboo, hit the internodes out. Slightly bend the bamboo inward by curving the bamboo. The nodes should split flat easily.

4 Weave the flattened bamboo for blinds or flooring.

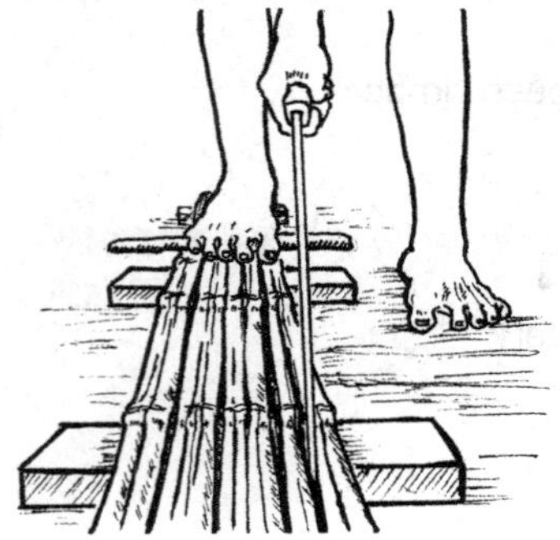

Sago palms

Sago palms grow well in the swamps and near rivers in the lowlands. The inner **fibres** are beaten to extract starchy substance for food. The sago palm stems and leaves are used commonly for building in traditional houses.

HOW TO PREPARE MOROTA
(sewn sago palm leaves)

1 Select and cut mature sago palm leaves.
2 Strip the leaf material off the midrib.
3 Sew sago leaves onto bamboo splints firmly with cane rope or strong bush rope
4 Leave the sewn leaves in the sun to dry.

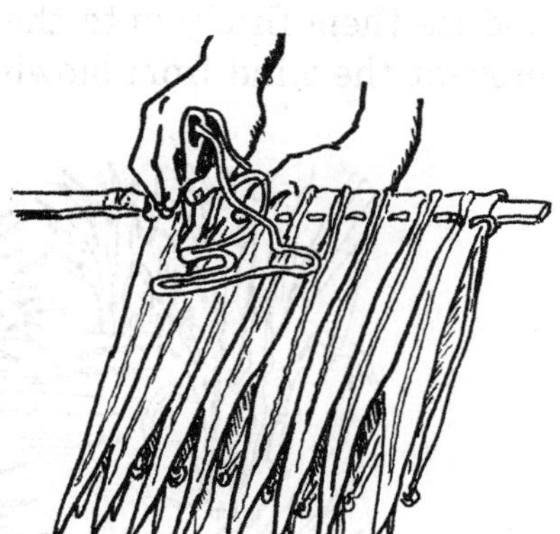

HOW TO PREPARE WEAVE MOROTA
(woven sago palm leaves)

1 Select and cut mature sago palm leaves. Do not remove the leaves from the midrib.
2 Weave the leaves together while the leaves are still fresh.
3 Leave the woven leaves to dry in the sun.

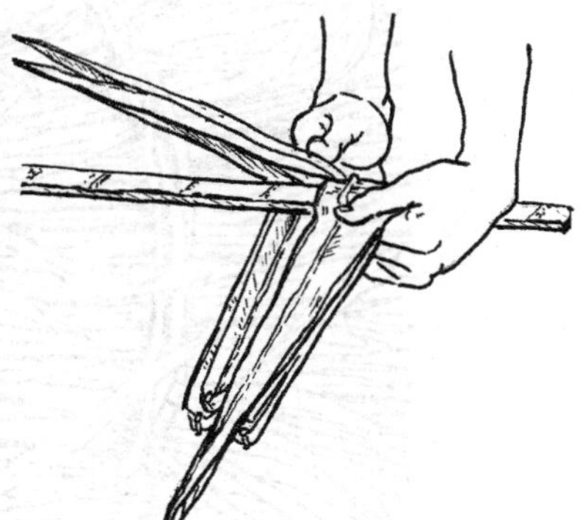

HOW TO PREPARE LIMBUNG (split palm)
WALLING AND FLOORING

1 Select and cut down mature palm tree.
2 Cut off the crown of the palm.
3 Split open the palm stem and remove the inner fibres.
4 Leave them to dry in the sun.

Kunai grasses

Kunai grasses, which are used for roofing, grow less than two metres tall. They grow well both in the highlands and lowlands. Nearly all houses in the rural areas of the highlands in Papua New Guinea have kunai grass roofs. You can use a bush knife to cut the grass or you can simply pull kunai grasses out. You then need to bundle them into small handfuls and tie them firmly onto the roof. That will hold the grasses firmly and prevent the wind from blowing them off the roofs.

Bush rope

There are many different species of vines in the forest, all of which can be used to make bush rope for tying together house posts, walls, roofs and floors. This rope is very strong and is a reliable method for fastening building materials. The methods of making the rope are very old, so older people can be helpful when you are learning how to make it. Bush cane can also be used for making rope.

HOW TO PREPARE BUSH CANES FOR ROPE

1. Identify and cut mature bush canes.
2. Split the canes into splints
3. Remove the inner fibres from the skin. The inner fibres can be used for tying when still wet.
4. You can dry the cane splints or keep them in the shade so that you can use them when they are still fresh and wet.

Pitpit

Beautiful traditional blinds are hand-woven and have various distinctive designs. Blinds can be used for both inside and outside walls. It is a good idea to paint them with varnish. Blinds used on external walls exposed to sun and rain will lose their colour and deteriorate rapidly unless they are protected by roof overhangs, which are low and big enough to protect them from rain and direct sunlight.

In the highlands of Papua New Guinea, people weave blinds from flattened pitpit or bamboo and sell them at the roadside. You can also weave blinds and sell them to other people.

There are three types of pipit. One kind which is not good for weaving blinds is found in both the highlands and the lowlands. The other two kinds of pitpit grow in the highlands. One is a soft pitpit commonly found growing in muddy swamps. The other type grows in the bush and along mountainsides. Both can be used for weaving blinds.

HOW TO PREPARE BLINDS

1 Select and cut matured pitpits.
2 Beat and flatten pitpits with a piece of timber.
3 Open the flattened pitpits, bend them inward and remove the dead leaf covering of the pitpits.
4 Weaving the blind mat on the ground. There are many different patterns and designs that you can use.

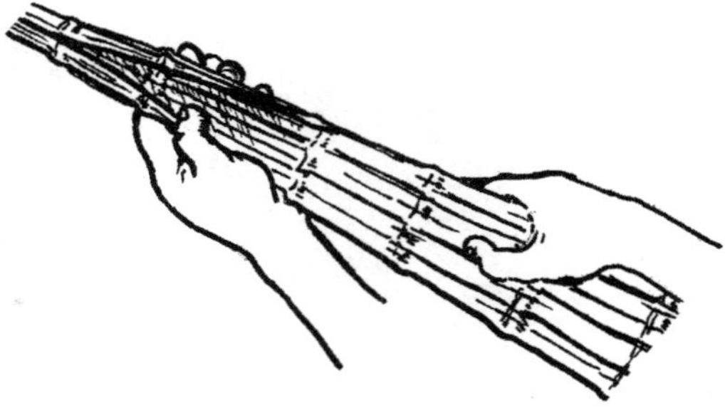

Bush timber

Trees from the bush are cut down for timber to build houses. Local people know which trees are suitable for building. They know the strength and the weakness of trees growing in their areas. Different types and sizes of timber can be used for different purpose in building. If you cannot identify and select suitable timber for building houses, your house will not last long. This is why it is important to ask local elders for their advice.

HOW TO PREPARE TIMBER

1 Identify and select the type of bush timber commonly used for building house in your local community
2 After felling the tree, remove its bark and let the tree dry.
3 Split big timber into suitable sizes and shapes. Leave these split pieces to dry
4 Treat the timber to prepare for using it for building. Usually, leaving the wood out in the sun to dry will make it suitable for building. This treatment also prevents termites from eating away at the timber.

Sawn timber

Sawn timber is a widely used building material in Papua New Guinea. Timbers are cut from mature trees with chainsaws or **walkabout sawmills**. Different types and sizes of timbers are cut for all kinds of works in building. Saw timber is used for all the framing, wall **cladding**, and flooring of buildings. It is a good idea to treat or seasoned sawn timbers well before using them.

Round post and poles

Traditional houses in Papua New Guinea use round posts and poles that are embedded in the ground. Joints are lashed to them with cane or strong bush rope. Round posts and poles are much cheaper and stronger than sawn timber and can be found anyplace where there is bush. Only simple tools—such as a bush knife and an axe—are required to work on them. Round posts and poles are ideal for many rural buildings such as classrooms, community centres and workshops. These posts and poles are often carved.

Properly treated posts are much cheaper than **galvanised** iron posts and will last just as long, if not longer. If not treated well, posts will be attacked by white ants and borers, which will lead to the quick decay of the building.

Basic building tool kit

Basic building tool kits can cost a lot of money. Although you do not need to buy and have all the tools, there are some essential tools you will need. But it is essential to have basic tools. The table below shows the basic tools you should have and their use.

Bush knife

Used for cutting timber and clearing bushes.

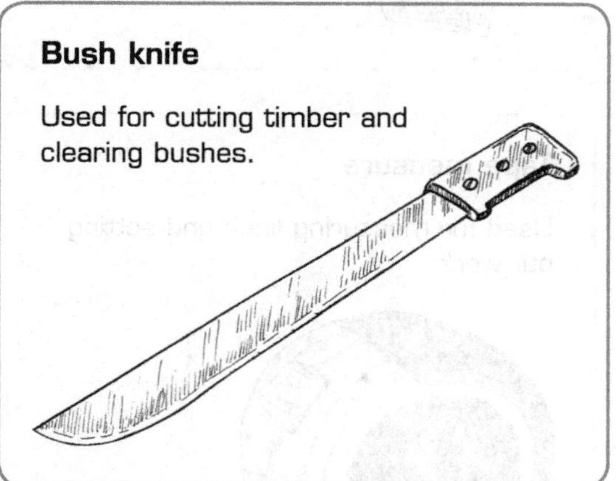

Axe

Used for felling trees and for cutting and splitting logs.

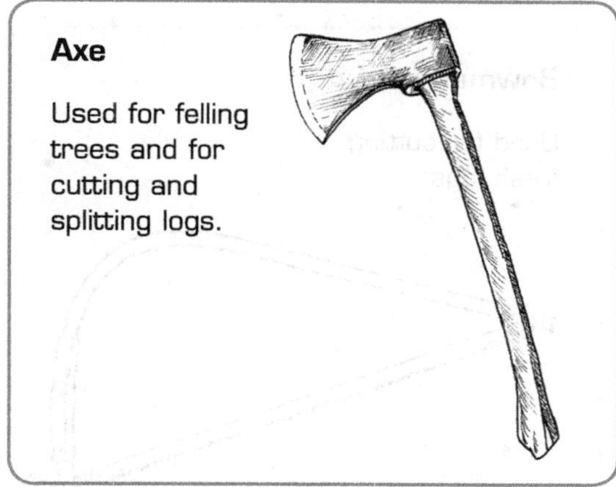

Spade and shovel

A spade is used for digging and levelling soil at a building site. A shovel is used for shifting gravel and concrete mixing.

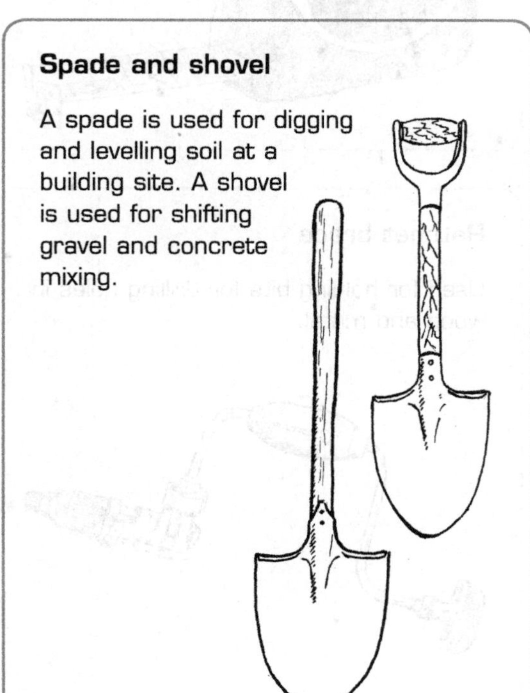

Claw hammer

Used for hitting nails into wood and pulling nails out of wood.

Crosscut saw

Used for cutting across timber.

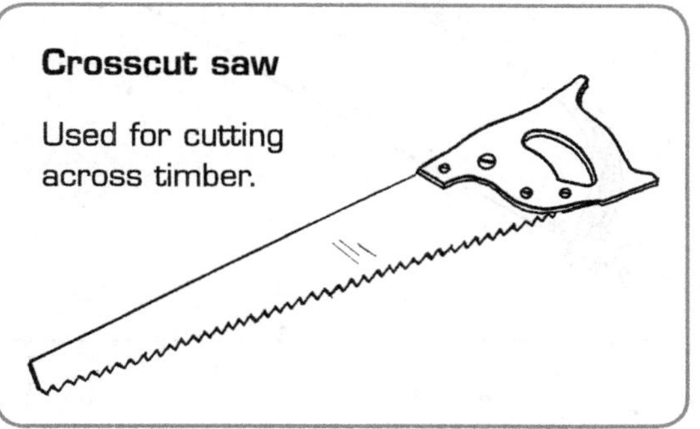

Bushman saw

Used for ripping timber.

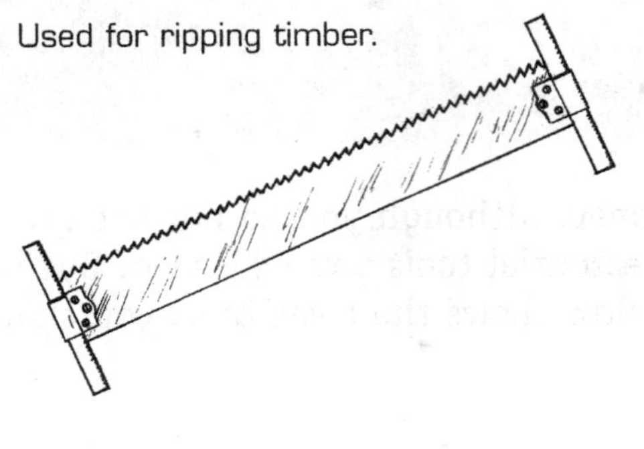

Bowman saw

Used for cutting fresh logs.

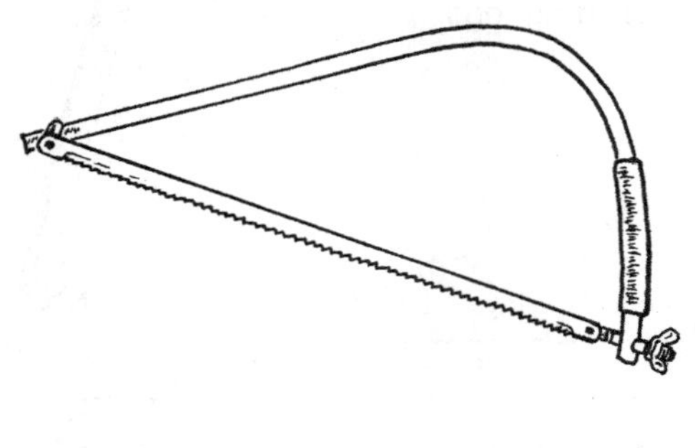

Hacksaw

Used for cutting metals. Has a number of changeable blades.

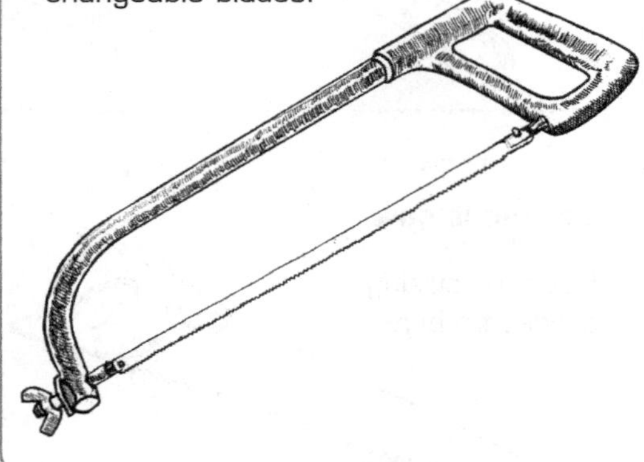

Try square

Used for marking out straight lines across timber, checking 90 degree angles and testing work.

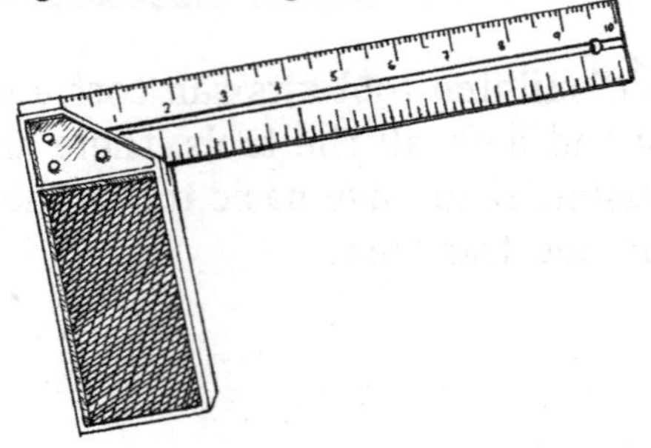

Tape measure

Used for measuring lines and setting out work.

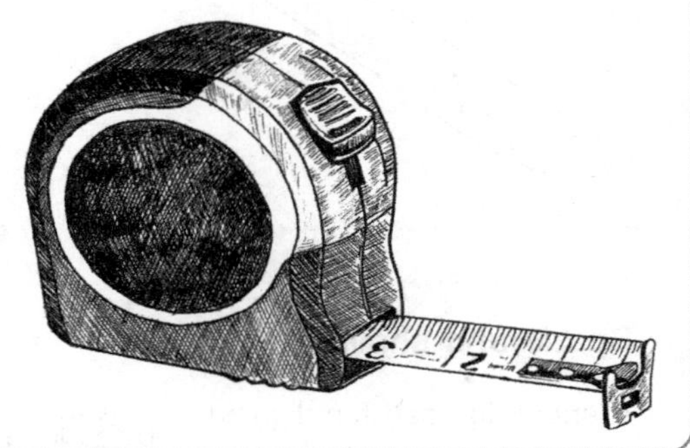

Ratchet brace

Used for holding bits for drilling holes in wood and metal.

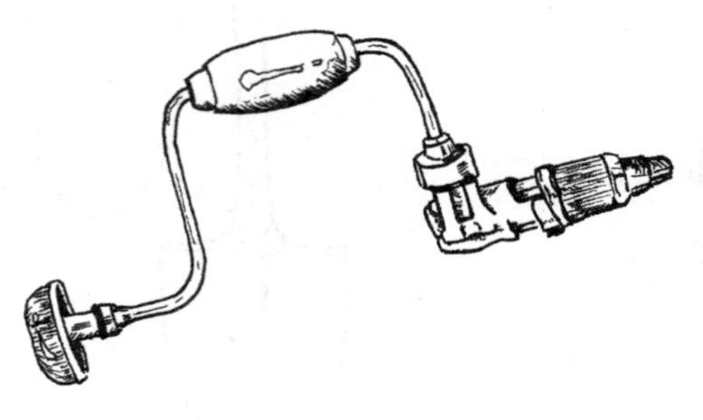

Spirit level

Used for testing the accuracy of level surfaces. It has a bubble of air floating in liquid. When the bubbles come to rest between two lines, it means that the surface is level.

Pencil

Used for marking and setting lines on prepared timber.

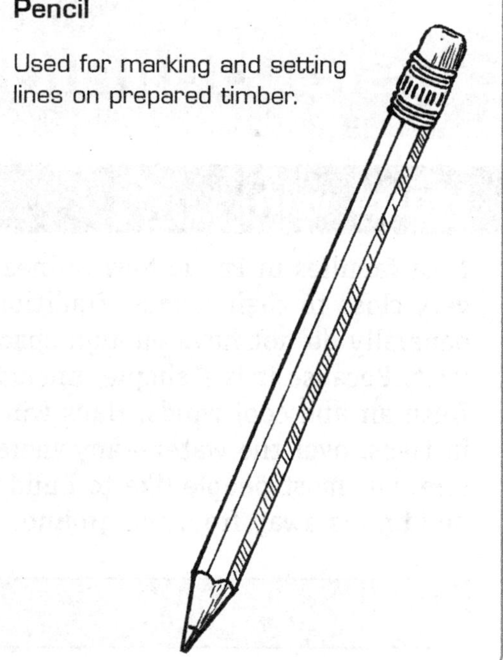

String

Used to mark diagonal and horizontal lines.

Basic buildings

Haus win

Most families in Papua New Guinea would like to have their haus win very close to their homes. Traditional family houses in Papua New Guinea generally do not have enough space to allow large families to meet and visit. Because it is a simple, unenclosed building, a haus win allows lots of fresh air and cool winds. Haus win can be built on the ground, on posts, in trees, over the water—anywhere you like. They can be any shape or size, but most people like to build their haus win on posts so that the building is away from the ground.

What you will need

- bush timber posts or milled timber posts
- bush timber
- strong bamboos
- corner poles
- palm stems
- sewn or woven sago palm leaves
- dried sago palm.

STEPS IN BUILDING

1 Clear the building site and mark out the building.

2 Dig the postholes and erect the posts firmly. The corner poles should be higher to hold the roofing.

3 Nail or fasten the **floor bearers** to the posts.

4 Fasten the **floor joists**.

Haus win *continued ...*

5

Lay the wall joists, nailing the top and bottom wall joists to the corner poles.

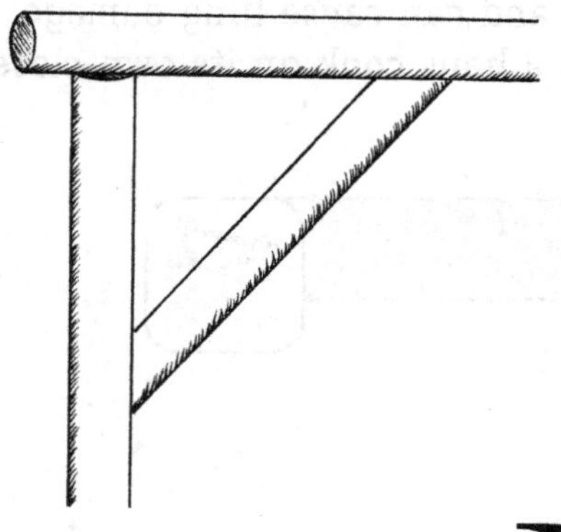

6

Lay a long timber over the two roof support poles.

7

Put up roof rafters over the roof support and fasten firmly.

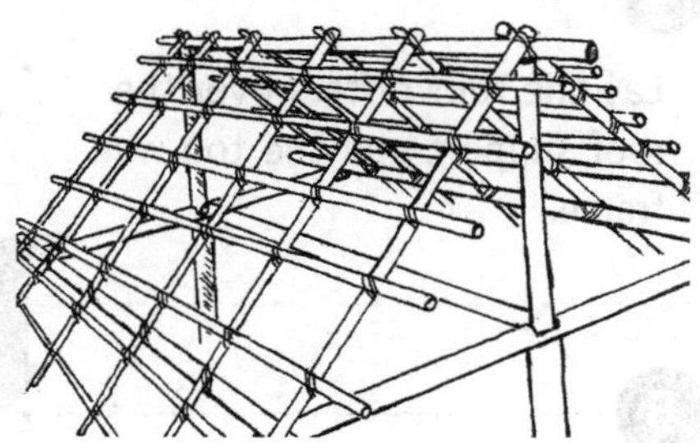

8

Tighten the sewn or woven morota onto the roof frame. Start from the bottom up. You can also use kunai instead of sago palm leaves.

9

Nail or fasten the palm stem or flattened bamboo onto the floor joist for flooring.

10

To prevent children from falling from the floor, erect a low wall using sago or bamboo splints.

Haus cook

A haus cook is a smaller house than the sleeping house. It is used for storing and cooking food in.

Most families in Papua New Guinea in rural villages cook food inside their sleeping houses. Many of these houses are small and so are often crowded. In many highlands houses, you will see roofs blackened by smoke from daily cooking. Smoke often fills the rooms and blocks the circulation of fresh air. This is unhygienic and can cause lung damage. That is why it is a very good idea to build a haus cook on its own close to your sleeping house.

What you will need

- bush timber posts
- bush timbers or milled timber
- nails (flat head, 3-4 inch)
- strong and matured bamboos
- strong sewn or woven sago palm leaves or kunai grass
- woven pitpit/bamboo blinds
- other suitable building materials.

STEPS IN BUILDING

1. Select and clear a building site close to your sleeping house.
2. Mark out and erect the corner and centre posts. Remember that all haus cooks are built on the ground floor. For houses built over the water or sea, the idea place for cooking is the veranda.
3. Lay across and fasten the roof support over the wall frame.
4. Lay the roof rafters over the roof support and the top wall frame.
5. Fasten sewn/woven morota or grass over the roof frame.

Haus cook *continued ...*

6

Nail woven blinds over the wall frame. Note: It is a good idea to build windows or leave open spaces for the free flow of air.

7

Make shelves and cupboards to store cooking utensils and food.

8

Make a fireplace on the ground for cooking food.

9

Build a door and hinge the door onto the house.

10

Build a small platform beside the cook for washing and drying cooking utensils.

Sun shelters and stalls

Sun shelters are temporary shelters build quickly with fewer bush materials. Such shelters are used for very short periods of time. Sun shelters are built for gatherings such as school fetes and sports carnivals. These types of building can also be used as stalls for selling food during such gatherings. The size of the shelter can be big or small depending or the number of people it will provide shelter. Not only will the shelter provide shelter from sun but also from rain.

What you will need

- raw bush timber
- strong bamboos
- strong bush vines
- coconut leaves
- kunai grass
- other bush grasses that can be used for roof covering.

STEPS IN BUILDING

1. Clear the building site.

2. Mark out the building outline and the post holes.

3. Plant the posts firmly into the post holes and reinforced the post firmly by packing the soil at the side of the posts. Both bamboos and timbers can be either used as posts. The number of rows required will depend on the size of the building.

4. Lay long bamboo or timber the top of the posts as rafter supports on all sides.

Sun shelters and stalls *continued …*

5

Lay timbers and bamboos across the roof supports as rafters and fasten them firmly.

6

Throw leaves and grasses over the rafters as roof covering. Palm leaves or coconut leaves can be woven and laid over or laid over without been woven.

7

Tidy over hangs of leaves neatly.

Roadside market houses

While many rural village people bring their local produces and artefacts to the markets to sell, it is now also common for people to sell their produce from roadside market houses. Fresh foods from the garden and cooked food are most commonly sold in these houses.

What you will need

- bush timbers and posts
- strong bush ropes
- nails (flat head, 3-4 inches nails)
- strong bamboos
- strong pitpit
- strong sewn or woven sago palm leaves or kunai grass.

STEPS IN BUILDING

1 Select a clear building site near the road. The site should be 4 to 6 metres away from the edge of the road. Do not build your house at the corner turn of the road.

2 Set out the outline of the building. If the roadside market house is to be used by many families, the building must be long and laid out in sections or rooms.

3 Erect posts to hold up the walls and roof.

4 Fasten the roof support to the wall frame.

5 Lay roof rafters over the support and fasten them tightly.

6 Throw woven sago palm leaves or kunai for roof covering and fasten them with bush ropes.

Roadside market houses *continued ...*

7

Enclose the back and the sides of the house. Leave a doorway in the front wall.

8

Make a shelf for selling items at the front side of the building facing the road.

9

Make a fireplace inside the building for cooking food for selling.

10

Cut off the overhanging leaves and timber.

Pit toilets

Pit toilets are often used in remote villages in Papua New Guinea. These are made by digging a deep pit in the ground. A floor with a centre hole is built over the pit. A seat with a lid sits over the central hole. A small house is built around the toilet. Pit toilets are cheap and easy to build. They can be made from either traditional or modern materials.

What you will need

- bush timber
- woven blinds
- bush ropes
- kunai grass / morota
- a flat piece of thick plywood with a central hole
- other suitable local materials.

STEPS IN BUILDING

1 Clear the building site, which should be about 8 to 10 metres away from the family house.

2 Dig the pit into the ground at a desired depth. The pit must be dug before you build the house over it.

3 Nail four pieces of timber together to form a square frame. You will nail the toilet seat to this frame.

Pit toilets *continued ...*

4

To make the toilet seat, cut a centre hole in the piece of flat plywood and lay it over the frame you made (Step 3). You could also build a box to put over the frame so that you will have a raised seat.

5

Place an empty kerosene drum in the centre as the centre hole.

6

Plant the posts around the hole.

7

Erect the roof frame.

8

Fasten the roof thatch firmly using grass or morota

9

Attached the woven blind around the wall frame

10

Make a good pathway from the house to the toilet.

Toilets over water

Toilets are built over water in most coastal villages. A walkway on posts is built from the shore to a distance of 30 metres or more over the water. At the end of the walkway there is a hole in the flooring. A small house is built around the toilet seats. These toilets are cheap and easy to build and the water carries the body wastes away. It is important that the walkway is strong and safe for children and elderly people to walk along.

What you will need

- bush timber posts
- mangrove posts
- milled timber
- strong bamboos
- strong bush rope
- nails
- split palm stems for walls
- flat milled timbers for walkway
- woven or sewn sago palm leaves.

STEPS IN BUILDING

1. Select a toilet site and dig and plant posts when the tide is low or the shore is dry.
2. Built a walkway to the toilet.
3. Build a platform where the actual toilet will be.
4. Erect the wall frame.
5. Erect the roof frame over the wall frame.

Toilets over water *continued ...*

6

Fasten the roof thatch with morota.

7

Nail or fasten splinted palm stems or palm leaves for walling. Leave small spaces in between walls so that sea breezes can flow through.

8

Make a seat with one or more holes over the hole in the floor.

Outside bathroom houses

A room for washing your body in private is called a bathroom. Traditionally, people of Papua New Guinea washed in the ocean or in creeks and rivers. We need to wash daily to feel clean and to remove sweat, dust, dirt and germs from our skin. People often choose to have a bathroom near their house because it is convenient and private. Most modem houses have bathrooms inside. If you are not living in a modern house, you need to build an outside bathroom.

What you will need

- bush timbers
- bamboos
- strong bush ropes
- river stones
- sewn or woven sago palm leaves or kunai grass
- split palm stems or strong pipit

STEPS IN BUILDING

1. Select and clear the building site a few metres from the house.
2. Set up the corner posts of the building.
3. Fasten the top wall frame timber to the posts.
4. Nail or fasten walling around the wall frame with strong pipit or split
5. Lay a piece of timber across the room to hang the water bucket and to hang wet towels to dry.
6. Lay river stones under the shower to prevent the area from becoming muddy.

Outside bathroom houses *continued …*

7

Dig a drainage ditch so that water can flow out of the bathroom.

8

If you want a roof, erect the roof rafters over the wall frame.

9

Fasten morota or kunai thatch over the roof.

10

Leave spaces between the roof and the wall so that the sun will shine in and dry the area, and so that fresh breezes can flow through the building.

Building traditional houses

Highlands roundhouses

Highlands roundhouses are usually built on the ground. Because it is cold at night, the floors are dug slightly into the ground so that the cold air does not blow through the house. There are no windows and only one door. There is usually a fireplace in the centre of the house. Beds are made all around the house. Cardboard boxes are used for storing food and utensils.

What you will need

- split bush timber
- bush timber poles
- milled timber
- strong and soft bamboos
- bush vines and nails grass.

STEPS IN BUILDING

1. Clear the building site.
2. Dig about 30 cm into the ground and remove loose soil till you reach a hard surface. Level the floor base so that it is even.
3. Locate the centre and dig a hole in which you will place the centre pole of the house.
4. Plant outside posts around in circle of the house at a wall height, and tie the top ends of the posts together with bent thin bamboo or cane.

Highlands roundhouses *continued ...*

5

Fasten woven blinds to the posts to create walls both inside and outside. Bamboo blinds are suitable for outside walls.

6

Nail two timbers across the wall posts to hold the centre post firmly.

7

Make a smaller platform at the top end of the centre roof post as a rest for the rafters.

8

Put up the roof rafters, spacing them evenly, from the cross timbers to the central post.

9

Tie strong or split bamboo around over the rafters with strong bush rope.

10

Bundle kunai grass and tie it to the rafters to make the roof.

11

Hang the door to the entrance.

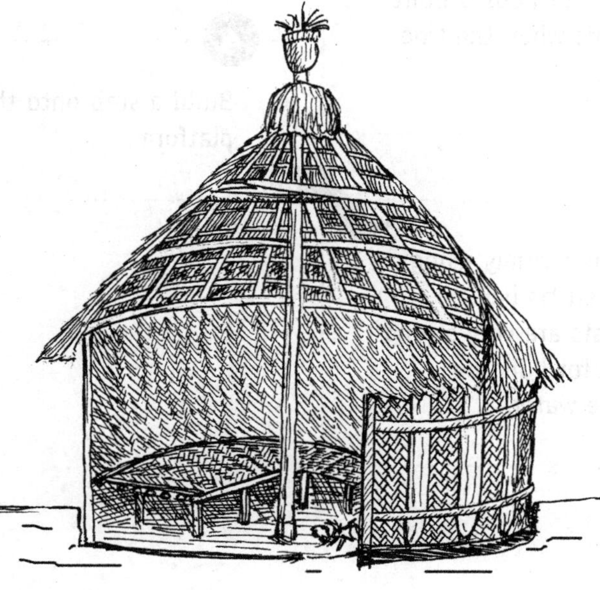

Coastal houses

The style and shape of houses may vary from one coastal area to another. Houses can be built on the sea, rivers, land and even up in the trees. All houses are built above the ground on posts. Most coastal houses have windows and have light walls to let fresh air flow through the house. The roofs can be made of either kunai thatch or morota thatch from sago palm leaves.

What you will need

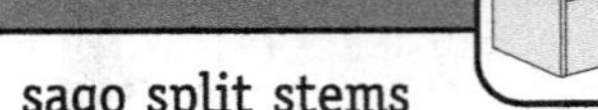

- prepared bush timber
- bush timber poles
- milled timber
- bamboos
- split palm stems
- sago leaves
- sago split stems
- grass
- bush canes
- bush vines
- nails.

STEPS IN BUILDING

1

Clear the site and mark out the floor of the house. Mark out the posts area and dig holes for the posts. For houses built on sea, do this when the tide is low.

2

Plant your posts firmly using strong prepared timber. Mangrove posts are very strong and are ideal for building houses on the water.

3

Fasten the floor platform firmly by nailing the floor bearers to the posts and floor joist timbers onto the floor bearers.

4

Build a step onto the floor platform.

Coastal houses *continued …*

5

Elect walling posts or studs and tie or nail them firmly. Lift the woven blind and attach it firmly onto the wall studs. If you are using sago palm or limbung for walling, tie it firmly with rattan cane or strong bush robe. (This step can be done after the roof is put up, as this will provide shade while you are working.)

6

Put up the roof posts and attach the roof rafters firmly.

7

On the top of the rafters, tie strong or split bamboo firmly. You will fasten the morota to this.

8

Tie the morota firmly, starting from the bottom up to the top of the roof.

9

Make the door for the entrance of the building. Hang it using hinges and staples.

Houses on the water

In many coastal areas of Papua New Guinea, houses are built on the water or sea. People living along the Sepik River build their houses on the banks of the river. Motuan villages in Koki and Hanuabada are built on the sea.

Manus people living along the coast also built theirs houses on the sea. These people connect their houses to the shore or each other's houses by building platforms up on stilts. They also use their canoes to travel from their houses to other places.

What you will need

- prepared mangrove timber posts
- bush timber
- milled timber
- split palm stems
- strong bush vines and cane ropes
- bamboo
- sago palm
- sewn sago leaves (morota) or grass.

STEPS IN BUILDING

1. Collect and prepare all building materials on the land.
2. Mark out the building site and dig holes for the post when there is a low tide.
3. Build a platform as the floor of the house.
4. Erect the wall frame of the house. You need to leave spaces for windows.
5. Build the roof frame over the wall frame.

Houses on the water *continued …*

6 Tie the prepared morota or kunai grass for roof covering.

7 Put up the wall blind or sago palm panelling over the wall frame.

8 Complete the flooring with split palm stems or flattened bamboo.

9 Hang the door to the entrance using hinges and staples.

Safety and first aid

It is important that anyone doing maintenance and repairs is familiar with basic first aid procedures, as accidents may happen. Every person should also have a basic first aid kit. The most common injuries to maintenance workers are caused by using tools carelessly or inappropriately.

These injuries can include:

- wounds
- excessive bleeding from wounds
- eye injuries
- broken bones.

Some basic first aid procedures

Wounds

A wound is a break in the skin caused by an injury. Cuts, grazes and burns are all wounds. Wounds are dangerous because they can cause bleeding and infection. Infection can be prevented by proper treatment of wounds.

What to do

- Stop any bleeding quickly (see the section on bleeding for more information).
- Small cuts and grazes may be cleaned with soap and water or an antiseptic solution before a clean dressing such as a clean towel, sheet or laplap is applied.

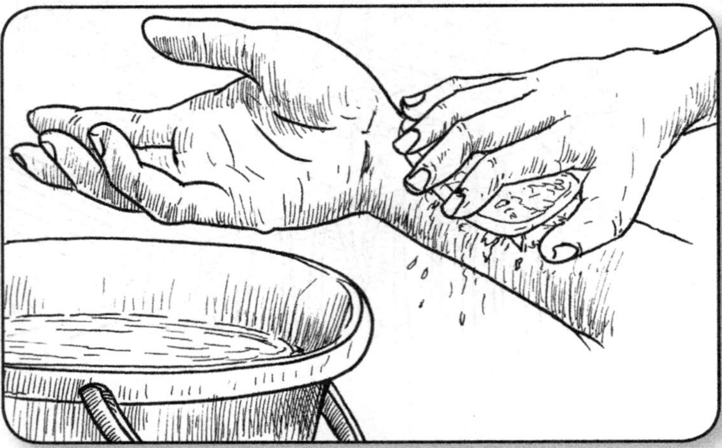

- Do not clean a wound that is bleeding heavily. Stop the bleeding if you can and take the victim to a health facility.

- Use plastic bags on your hands to stop you coming into contact with any blood. Dispose of plastic and any soiled dressings into a sealed container, and either burn it or bury it well.

- A puncture wound, such as an injury from a rusty nail, may become infected later. It is wise to see a health worker, because an anti-tetanus injection may be required.

Bleeding

If too much blood is lost from the body through a wound or injury, the victim may become seriously ill and even die. Always stop the bleeding as soon as possible.

What to do

- Stop the flow of blood by pressing the skin edges together and placing a bulky dressing, such as a clean towel, sheet or laplap, firmly on the wound.

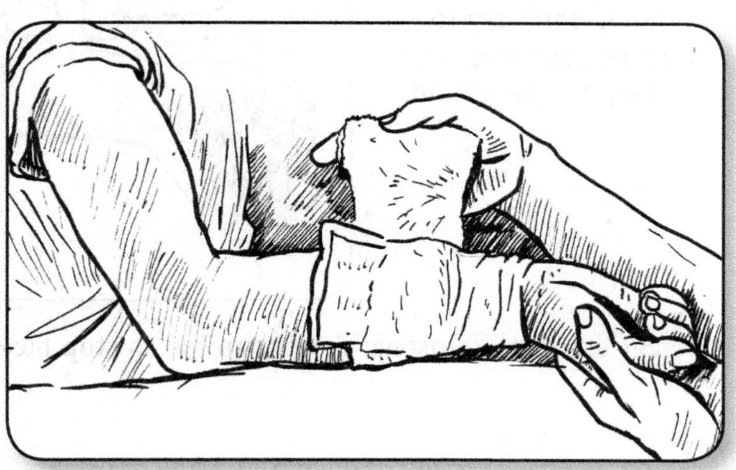

- Keep firm pressure on the wound for ten minutes, with the victim in a resting position. If the wound is on a limb, the limb should be raised to slow down the blood flow.
- If bleeding continues through the first dressing, put another dressing over the first dressing. Do *not* remove the first one.
- Keep the injured part raised for a while after the bleeding has stopped. A health worker should check severe wounds.

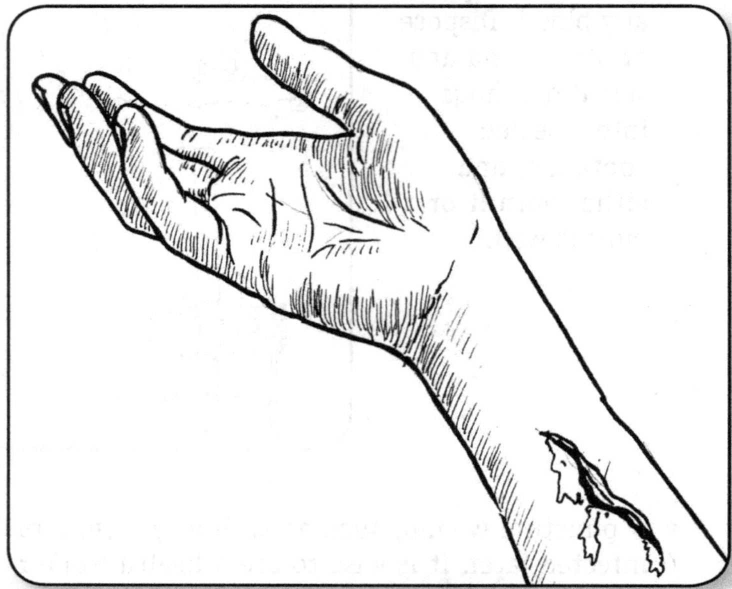

- Foreign bodies such as glass, metal or splinters can be removed if they can be seen clearly on the surface of the wound. Don't try to remove anything that is deep in the wound.

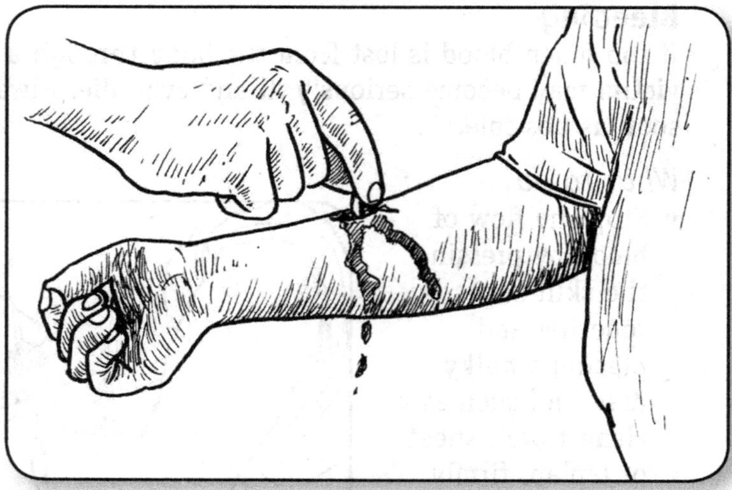

- Never use dirt, kerosene, lime or coffee to stop bleeding.

Eye injuries

Great care is needed when a foreign body gets into the eye. Rough treatment may lead to permanent loss of vision.

What to do

- Help the victim flush the eye under cool, running water. This may dislodge the foreign body.

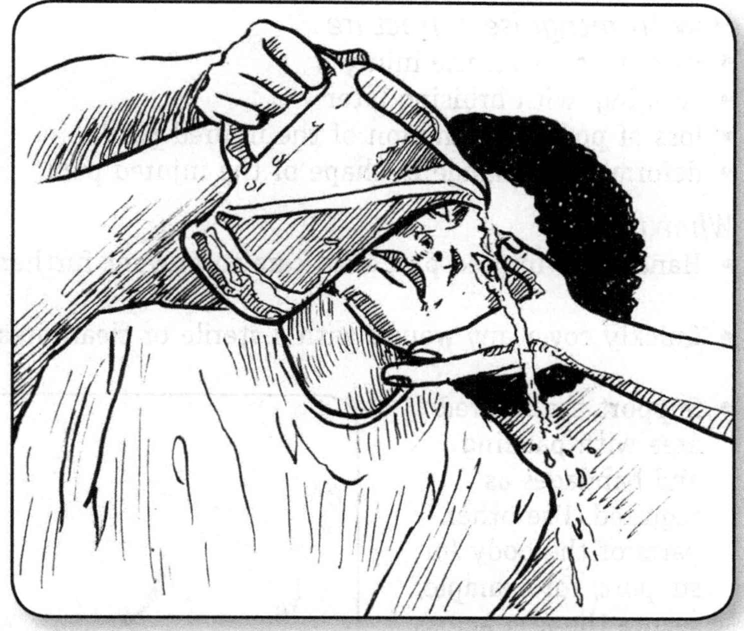

- If this is unsuccessful, gently pull down the lower eyelid and look for the object.
- If it can be seen on the eyelid, or on the white part of the eye, gently remove it with a corner of a clean handkerchief or tissue. *Never* use sharp objects to remove the foreign body.

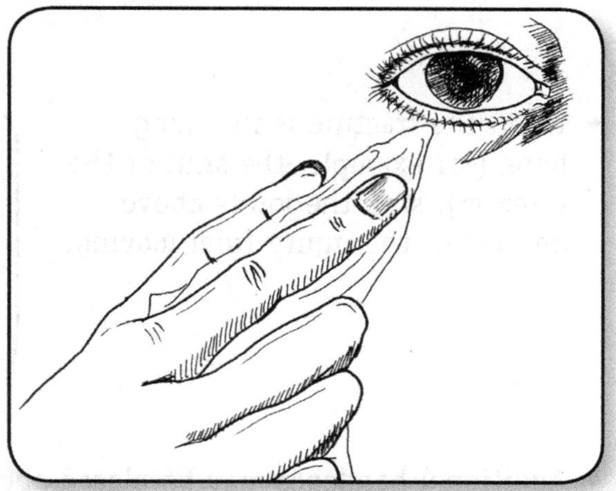

- If the object cannot be seen or is present over the coloured part of the eye, *do not try to remove it*. Cover the eye with a clean pad and seek medical aid.

Broken bones (fractures)

A cracked or broken bone is called a fracture. It is often hard to tell whether or not a bone is broken. When there is doubt, the injury must be treated as a fracture.

How to recognise a fracture

- pain at, or near, the injury
- swelling, with bruising later
- loss of power or function of the injured part
- deformity or change in shape of the injured part.

What to do

- Handle the injured part with care to prevent further pain and shock.
- Quickly cover any wound with a sterile or clean dressing.
- Support the injured area with padding and bandages as required. Use other parts of the body for support; for example, secure the arm across the chest to support broken ribs.

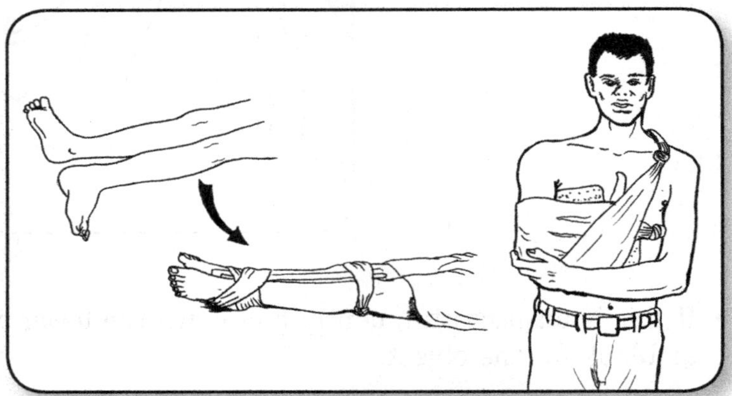

- When the fracture is in a long bone (for example, the shin or the forearm), stop the joints above and below the injury from moving.

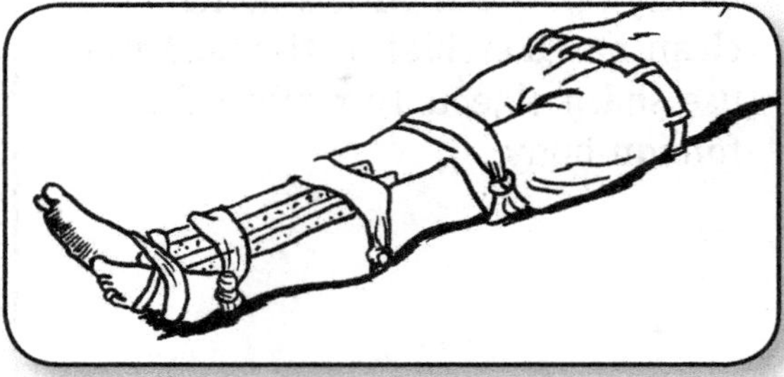

- Additional bandages may be placed above or below the injury but these should never be placed over the fracture.
- Seek medical aid as soon as possible because special treatment may be required.

Glossary

bamboo	a tall bush plant with hard hollow stem
blinds	a wall cover to cover the building framework woven from flattened bamboos
cladding	the walls of buildings
fibre	very thin thread substance in the inner part of the palm tree
floor bearer	timber fastened to the posts to carry the load of the floor
floor joist	timber that runs over the bearers and supports the flooring
galvanise	to coat metal with zinc to protect metal from rusting
internode	the hollow space in between the bamboo nodes
grass	a grass that is used for roof covering on Papua New Guinea village houses
kwila	a tree which last for a very long time and commonly used as posts for houses
luesino	a legume tree plant among cocoa trees which is commonly used as house posts by Tolai people
mangrove	a coastal tree with long roots that hang over the sea and penetrate deep into the sea floor; mangrove posts are very strong and can withstand the effect of sea water over a long period of time
morota	sewn or woven sago leaves for roof covering for Papua New Guinea village houses
node	the hard partition in between the hollow of bamboo
overhang	to lay across part of something
platform	the floor of the house

pitpit	bush material used for weaving blinds for houses
rafters	timber that carries the roof covering
sago palm	a palm tree whose stem and leaves are commonly used for building Papua New Guinea village houses
season	treat raw cut timber (air seasoning is the common method used to treat timber in Papua New Guinea)
tar	a black sticky substance used on roads or patching holes in the corrugated iron
Tolai	name given the people of East New Britain province, especially people living along the coast
tolas	a strong tree used as house posts by people of Bougainville Island
wakabout sawmill	portable sawmill for sawing timber
weatherboard	moulded board used for outside wall

Notes

Notes *continued …*